I0190642

DELIVERANCE

from

ASTRAL

PROJECTION

DR. TAQUETTA BAKER

Kingdom Wellness Counseling and Mentoring Center

Christian Counseling

Ebook Products

DELIVERANCE FROM ASTRAL PROJECTION

Dr. TAQUETTA BAKER

Amazon Publications
Muncie, IN

Kingdom Wellness Counseling and Mentoring Center
Learning Applicable Relationship Tools

This manual is based in over 22 years experience in professional counseling.

kingdomwellnesscenter@gmail.com

Kingdomshifters.com

KSWU.NET

Connect with Taquetta via Facebook or YouTube

Copyright 2023 – Kingdom Wellness Counseling & Mentor Center. All rights reserved.

Images are either copyright free, public domain images or used with permission of the graphic artist.

This book is protected by the copyright laws of the United States of America. This book may not be reprinted for commercial gain or profit. The use of occasional page copying for personal or group study is permitted and encouraged. Permission will be granted with written request.

TABLE OF CONTENTS

VISION STATEMENT

Kingdom Wellness offers a revolutionary theory of bridging mental and physical health with biblical truths, faith-based counseling, deliverance and healing principles. This is a holistic ideology of the total person – body, soul (mind, will, emotions), and spirit becoming one.

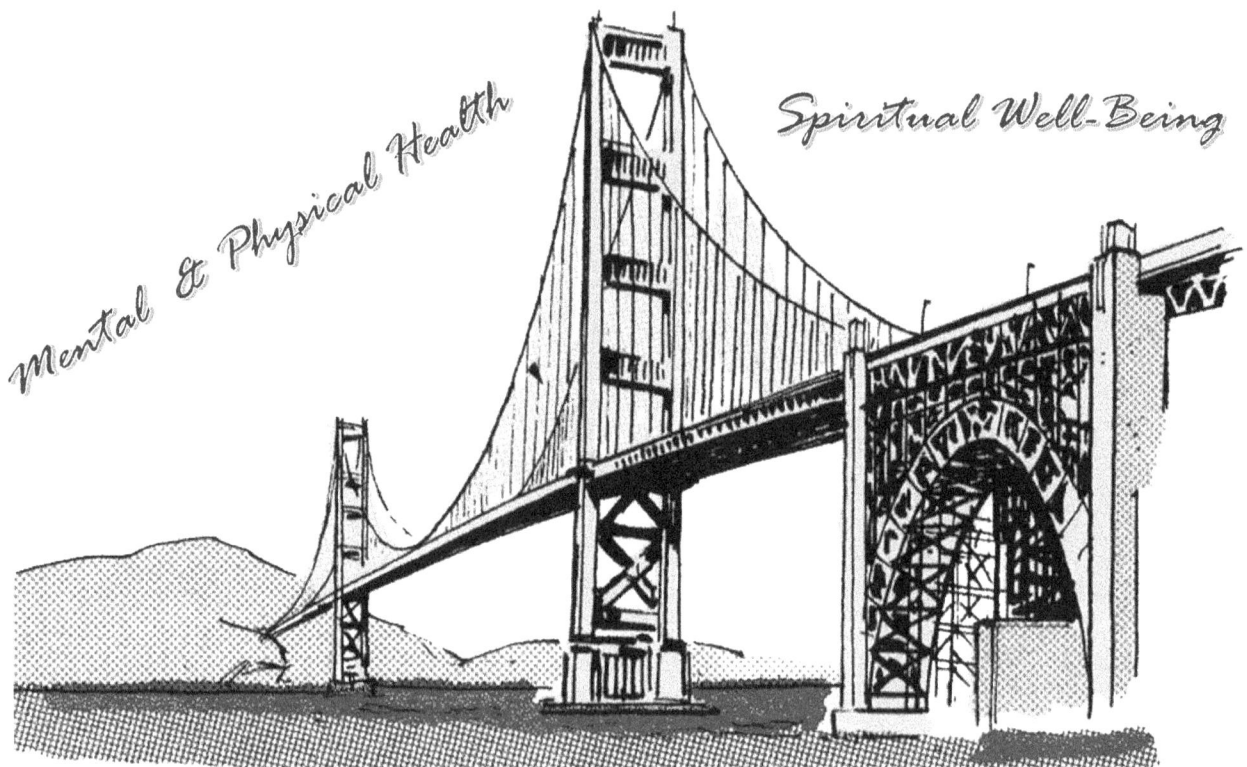

Kingdom Wellness

Mental & Physical Health

Spiritual Well-Being

"Psychological theories are valuable for guiding practice in education, mental health, business, and other domains. They provide answers to intrinsically interesting questions concerning many kinds of thinking including perception, emotion, learning, and problem-solving."[i]

TAQUETTA BAKER, FOUNDER

As an adopted child, Taquetta knows the lifelong journey of appreciating both spiritual and natural relationships and encounters. She understands being strengthened by the pain that life can bring while embracing the joy of unexpected connections. Taquetta has transformed her experiences to establish a legacy of pioneering and fearlessly leading others.

Taquetta Baker is the founder of Kingdom Shifters Ministries (KSM), Kingdom Shifters Empowerment Church, and Kingdom Wellness Counseling and Mentoring Center. She has authored over 50 books and two prayer CD's. Taquetta has a Doctorate in Ministry, a Master's Degree in Community Counseling with an emphasis on Marriage, Children and Family Counseling, a Bachelor's Degree in Psychology and Associates Degree in Business Administration. Taquetta has a Therapon Belief Therapist Certification from the Therapon Institute, which provides faith-based counseling and ministry training. She is a certified Leadership & Executive Life Coach and has written her own Kingdom Wellness Counseling & Life Coaching theory and curriculum.

Taquetta serves in the mental health field as a Behavioral Consultant. She enjoys working with individuals and families who experience a broad range of psychological, emotional, social, relational, and spiritual challenges. Her outreach demonstrates cultural agility across a spectrum of ages, ethnicities, and socio-economic backgrounds. She is committed to empowering others with launching ministries, businesses, and books. She provides mentoring, counseling and vision launching through her Kingdom Wellness Counseling and Mentoring Center. With over 22 years of faith-based and professional counseling experience, her reputation is one who transforms lives and families through balancing biblical principles with applicable tools and strategies.

Taquetta serves on the Board of Directors for New Day Community Ministries, Inc. and is a graduate of the Eagles Dance Institute under Dr. Pamela Hardy with a license in liturgical dance. Before pioneering her own ministry, Taquetta was a dedicated member of Christ Temple Global Ministries for 14 years. She served pioneered Shekinah Expressions dance ministry and served in the role of prophet, teacher, presbytery board member, and overseer of the Altar Workers Ministry. Taquetta receives mentoring and ministry covering from Bishop Jackie Green, Founder of JGM-National Prayer Life Institute (Phoenix, AZ), and was ordained as an Apostle on June 7, 2014.

The Bible is full of stories that are centered around digging or receiving from wells which represent stability and deep places of renewal. Taquetta flows through the spiritual wells of warfare, worship, counseling and deliverance. Taquetta's mantle is an apostolic directive of judging and establishing God's kingdom in people, ministries, communities, and regions. Taquetta travels in foreign missions and throughout the United States. She has mentored and established dance teams, altar workers, counseling programs, and deliverance and prophetic ministries. Taquetta ministers in the areas of fine arts, systems of prayer, fivefold ministry, deliverance, healing, miracles, atmospheric worship, and counseling. Her mission is to empower and train others to identify and embrace their destiny.

Shift!

DELIVERANCE FROM ASTRAL PROJECTION

> *In order to be delivered from astral projection, it is important to understand witchcraft terminology. I will share that terminology with you first and then share revelation regarding deliverance from astral projection.*

Regions & Spheres Of Influence – A Region is an area of space, land, or atmosphere. A sphere of influence is the place, environment, or region to which a person or thing exists, influences, serves, governs, or operates.

God has placed specific revelation and knowledge about our identity and calling into the regions and spheres of influence he has called us to live and govern in.

*Isaiah 49:1-4 Listen, O isles, (**isles means region in Hebrew**) unto me; and hearken, ye people, from afar; The Lord hath called me from the womb; from the bowels of my mother hath he made mention of my name. And he hath made my mouth like a sharp sword; in the shadow of his hand hath he hid me, and made me a polished shaft; in his quiver hath he hid me; And said unto me, Thou art my servant, O Israel, in whom I will be glorified. Then I said, I have laboured in vain, I have spent my strength for nought, and in vain: yet surely my judgment is with the Lord, and my work with my God.*

New Living Bible Listen to me, all you in distant lands! Pay attention, you who are far away! The Lord called me before my birth; from within the womb he called me by name. He made my words of judgment as sharp as a sword. He has hidden me in the shadow of his hand. I am like a sharp arrow in his quiver. He said to me, "You are my servant, Israel, and you will bring me glory." I replied, "But my work seems so useless! I have spent my strength for nothing and to no purpose. Yet I leave it all in the Lord's hand; I will trust God for my reward."

Many principalities, powers, territorial spirits, rulers in high places will contend with people who are seeking to govern or influence in their regions and spheres. Sometimes gateways, portals, bewitchment, demonic assignments, and attacks, will have to be closed and broken from or off people's regions and spheres of influence. Sometimes having the person decree and agree with what God has spoken to them concerning their region and spheres of influence will need to be released into the womb and atmosphere of the region, while cleansing negative words and witchcraft that has been released against the person.

Since a region is a space and can include the heavenlies, the atmosphere and the land, words and witchcraft practices may have to be cleansed and called out of the moon, stars, sun, elements, frequencies, airways, land, and the underworld. Sometimes witchcraft has to be broken from bodies of water such as oceans, rivers, lakes, and ponds.

Idols – Let's be clear, idols do not have any power. As people carve, identify, and worship idols, they receive their power from the demons that are attached to them. But the idol, the artifact, person, position, element, statue, carvings, images in and of themselves, do not have any power. They are empowered by the demons that inhabit them.

Jeremiah 10:8-11 But they are altogether brutish and foolish: the stock is a doctrine of vanities. Silver spread into plates is brought from Tarshish, and gold from Uphaz, the work of the workman, and of the hands of the founder: blue and purple is their clothing: they are all the work of cunning men. But the LORD is the true God, he is the living God, and an everlasting king: at his wrath the earth shall tremble, and the nations shall not be able to abide his indignation. Thus shall ye say unto them, the gods that have not made the heavens and the earth, even they shall perish from the earth, and from under these heavens.

Jeremiah 10:14-15 Every man is brutish in his knowledge: every founder is confounded by the graven image: for his molten image is falsehood, and there is no breath in them. They are vanity, and the work of errors: in the time of their visitation they shall perish.

Jeremiah 51:17-19 Every man is brutish by his knowledge; every founder is confounded by the graven image: for his molten image is falsehood, and there is no breath in them. They are vanity, the work of errors: in the time of their visitation they shall perish. The portion of Jacob is not like them; for he is the former of all things: and Israel is the rod of his inheritance: the LORD of hosts is his name.

Psalm 97:7-12 Confounded be all they that serve graven images, that boast themselves of idols: worship him, all ye gods. Zion heard, and was glad; and the daughters of Judah rejoiced because of thy judgments, O LORD. For thou, LORD, art high above all the earth: thou art exalted far above all gods. Ye that love the LORD, hate evil: he preserveth the souls of his saints; he delivereth them out of the hand of the wicked. Light is sown for the righteous, and gladness for the upright in heart. Rejoice in the LORD, ye righteous; and give thanks at the remembrance of his holiness.

Psalm 135:16-18 *The idols of the heathen are silver and gold, the work of men's hands. They have mouths, but they speak not; eyes have they, but they see not; They have ears, but they hear not; neither is there any breath in their mouths. They that make them are like unto them: so is every one that trusteth in them.*

Psalm 115:4-8 *Their idols are silver and gold, the work of men's hands. They have mouths, but they speak not: eyes have they, but they see not: They have ears, but they hear not: noses have they, but they smell not: They have hands, but they handle not: feet have they, but they walk not: neither speak they through their throat. They that make them are like unto them; so is every one that trusteth in them.*

Habbukak 2:18-19 *What profiteth the graven image that the maker thereof hath graven it; the molten image, and a teacher of lies, that the maker of his work trusteth therein, to make dumb idols? Woe unto him that saith to the wood, Awake; to the dumb stone, Arise, it shall teach! Behold, it is laid over with gold and silver, and there is no breath at all in the midst of it.*

Isaiah 42:17 *They shall be turned back, they shall be greatly ashamed, that trust in graven images, that say to the molten images, Ye are our gods.*

Isaiah 44:17-20 *And the residue thereof he maketh a god, even his graven image: he falleth down unto it, and worshippeth it, and prayeth unto it, and saith, Deliver me; for thou art my god. They have not known nor understood: for he hath shut their eyes, that they cannot see; and their hearts, that they cannot understand. And none considereth in his heart, neither is there knowledge nor understanding to say, I have burned part of it in the fire; yea, also I have baked bread upon the coals thereof; I have roasted flesh, and eaten it: and shall I make the residue thereof an abomination? shall I fall down to the stock of a tree? He feedeth on ashes: a deceived heart hath turned him aside, that he cannot deliver his soul, nor say, Is there not a lie in my right hand?*

Isaiah 45:14 *They shall be ashamed, and also confounded, all of them: they shall go to confusion together that are makers of idols.*

Romans 1:22-24 *Professing themselves to be wise, they became fools, And changed the glory of the uncorruptible God into an image made like to corruptible man, and to birds, and fourfooted beasts, and creeping things. Wherefore God also gave them up to uncleanness through the lusts of their own hearts, to dishonour their own bodies between themselves: Who changed the truth of*

God into a lie, and worshipped and served the creature more than the Creator, who is blessed for ever. Amen.

Acts 17:22-31 *Then Paul stood in the midst of Mars' hill, and said, Ye men of Athens, I perceive that in all things ye are too superstitious. For as I passed by, and beheld your devotions, I found an altar with this inscription, TO THE UNKNOWN GOD. Whom therefore ye ignorantly worship, him declare I unto you. God that made the world and all things therein, seeing that he is Lord of heaven and earth, dwelleth not in temples made with hands; Neither is worshipped with men's hands, as though he needed any thing, seeing he giveth to all life, and breath, and all things; And hath made of one blood all nations of men for to dwell on all the face of the earth, and hath determined the times before appointed, and the bounds of their habitation; That they should seek the Lord, if haply they might feel after him, and find him, though he be not far from every one of us: For in him we live, and move, and have our we are also his offspring. Forasmuch then as we are the offspring of God, we ought not to think that the Godhead is like unto gold, or silver, or stone, graven by art and man's device. And the times of this ignorance God winked at; but now commandeth all men every where to repent: Because he hath appointed a day, in the which he will judge the world in righteousness by that man whom he hath ordained; whereof he hath given assurance unto all men, in that he hath raised him from the dead.*

Dictionary.com defines idol as:
1. an image or other material object representing a deity to which religious worship is addressed
2. any person or thing regarded with blind admiration, adoration, or devotion
3. a mere image or semblance of something, visible but without substance, as a phantom
4. a figment of the mind; fantasy
5. a false conception or notion; fallacy

Idolatry is the serving of self and one's own ideologies. It results in invoking God's wrath, anger, and judgement. God is adamant about not being worshipped, identified, or compared to other gods. **HE NOT HAVING IT.**

Exodus 20:5 *Thou shalt not bow down thyself to them, nor serve them: for I the LORD thy God [am] a jealous God, visiting the iniquity of the fathers upon the children unto the third and fourth [generation] of them that hate me.*

Exodus 23:13 *And in all things that I have said unto you be circumspect: and make no mention of the name of other gods, neither let it be heard out of thy mouth.*

Deuteronomy 6:13-15 *"Thou shalt fear the LORD thy God, and serve him, and shalt swear by his name. Ye shall not go after other gods, of the gods of the people which are round about you;*

(For the LORD thy God is a jealous God among you) lest the anger of the LORD thy God be kindled against thee, and destroy thee from off the face of the earth.

Deuteronomy 8:19-20 *And it shall be, if thou do at all forget the LORD thy God, and walk after other gods, and serve them, and worship them, I testify against you this day that ye shall surely perish. As the nations which the LORD destroyeth before your face, so shall ye perish; because ye would not be obedient unto the voice of the LORD your God."*

Deuteronomy 11:16-17, 26-28 *Take heed to yourselves, that your heart be not deceived, and ye turn aside, and serve other gods, and worship them; And then the LORD'S wrath be kindled against you, and he shut up the heaven, that there be no rain, and that the land yield not her fruit; and lest ye perish quickly from off the good land which the LORD giveth you . . . Behold, I set before you this day a blessing and a curse; A blessing, if ye obey the commandments of the LORD your God, which I command you this day: And a curse, if ye will not obey the commandments of the LORD your God, but turn aside out of the way which I command you this day, to go after other gods, which ye have not known.*

Joshua 23:15-16 *Therefore it shall come to pass, that as all good things are come upon you, which the LORD your God promised you; so shall the LORD bring upon you all evil things, until he have destroyed you from off this good land which the LORD your God hath given you. When ye have transgressed the covenant of the LORD your God, which he commanded you, and have gone and served other gods, and bowed yourselves to them; then shall the anger of the LORD be kindled against you, and ye shall perish quickly from off the good land which he hath given unto you."*

Joshua 24:21-27 *And the people said unto Joshua, Nay; but we will serve the LORD. And Joshua said unto the people, Ye are witnesses against yourselves that ye have chosen you the LORD, to serve him. And they said, We are witnesses. Now therefore put away, said he, the strange gods which are among you, and incline your heart unto the LORD God of Israel. And the people said unto Joshua, The LORD our God will we serve, and his voice will we obey. So Joshua made a covenant with the people that day, and set them a statute and an ordinance in Shechem. And Joshua wrote these words in the book of the law of God, and took a great stone, and set it up there under an oak, that was by the sanctuary of the LORD. And Joshua said unto all the people, Behold, this stone shall be a witness unto us; for it hath heard all the words of the LORD which he spake unto us: it shall be therefore a witness unto you, lest ye deny your God.*

Romans 2:8 *But for those who are self-seeking and who reject the truth and follow evil, there will be wrath and anger.*

1John 2:15-16 Love not the world, neither the things that are in the world. If any man love the world, the love of the Father is not in him. For all that is in the world, the lust of the flesh, and the lust of the eyes, and the pride of life, is not of the Father, but is of the world.

1John 5:21 Little children, keep yourselves from idols. Amen.

It would be beneficial for the deliverance minister to study the book of **Exodus** and **Deuteronomy**. It is full of insight regarding God's hatred for idolatry.

People want deliverance but they do not want to give up their idols. People like to build idols because they want a god they can see and control. They want to be able to dictate to the god what they want and how the idol should be worshipped. What they do not realize is, once a demon inhabits that idol, they are drawn into worshipping that idol the way that demon and its master – SATAN – demands to be worshipped. SATAN WAS WORSHIP!

Ezekiel 20:14-19 Thou art the anointed cherub that covereth; and I have set thee so: thou wast upon the holy mountain of God; thou hast walked up and down in the midst of the stones of fire. Thou wast perfect in thy ways from the day that thou wast created, till iniquity was found in thee. By the multitude of thy merchandise they have filled the midst of thee with violence, and thou hast sinned: therefore I will cast thee as profane out of the mountain of God: and I will destroy thee, O covering cherub, from the midst of the stones of fire. Thine heart was lifted up because of thy beauty, thou hast corrupted thy wisdom by reason of thy brightness: I will cast thee to the ground, I will lay thee before kings, that they may behold thee. Thou hast defiled thy sanctuaries by the multitude of thine iniquities, by the iniquity of thy traffick; therefore will I bring forth a fire from the midst of thee, it shall devour thee, and I will bring thee to ashes upon the earth in the sight of all them that behold thee. All they that know thee among the people shall be astonished at thee: thou shalt be a terror, and never shalt thou be any more.

His very identity and physical being was worship. This was his position and existence in heaven. Satan has been damned to hell for wanting to be worshipped like God and so is the judgment of those who worship him and his idols.

Isaiah 14:12-21 How art thou fallen from heaven, O Lucifer, son of the morning! how art thou cut down to the ground, which didst weaken the nations! For thou hast said in thine heart, I will ascend into heaven, I will exalt my throne above the stars of God: I will sit also upon the mount of the congregation, in the sides of the north: I will ascend above the heights of the clouds; I will be like the most High. Yet thou shalt be brought down to hell, to the sides of the pit.

They that see thee shall narrowly look upon thee, and consider thee, saying, Is this the man that made the earth to tremble, that did shake kingdoms; destroyed the cities thereof; that opened not *the house of his prisoners? All the kings of the nations, even all of them, lie in glory, every one in his own house. But thou art cast out of thy grave like an abominable branch, and as the raiment of those that are slain, thrust through with a sword, that go down to the stones of the pit; as a carcase trodden under feet. Thou shalt not be joined with them in burial, because thou hast destroyed thy land, and slain thy people: the seed of evildoers shall never be renowned. Prepare slaughter for his children for the iniquity of their fathers; that they do not rise, nor possess the land, nor fill the face of the world with cities.*

Because the hooking of idolatry is so bewitching, enticing, and self-serving, people will need discipleship to sustain the deliverance from idols. They will have to be deprogrammed from the cultish mindsets they have yielded to. They will also need to learn how idolatry has operated in their family lines and the propensities they have for engaging in idolatry and being drawn to idolatrous movies, music, encounters, artifacts, and etc. Until they are discipled and fully understand and honor God's boundary line, they will continue to open the door to idolatry. They will keep needing deliverance because they have no TRUTH to sustain it. Also, those propensities have to be uprooted out of them and their generational line.

People also need to learn to love what God loves and hate what he hates. People love their sins. They dread relinquishing their pleasures. For this reason, many abstain but are not fully delivered. They are constantly in agony and have to choose righteousness, rather than being and living righteously, where they are not tormented by the choices they have to make for righteousness sake.

Many people serve God because they enjoy the benefits of his kingdom, but they do not love him or his kingdom. People love Satan's kingdom and long for what is in his kingdom, while claiming to be of God's kingdom. They need to be broken in their heart and will to love God and his kingdom. Only being discipled in the truth of God's word and principles can produce the fruit of love for God and what is of him. It would be beneficial to connect people to a discipleship class, a counselor or mentor who can disciple them for a season until they fully learn how to live in God and by the leading of the Holy Spirit. Teach them how to choose God where they have no other gods before him.

Matthew 5:18 *This people draweth nigh unto me with their mouth, and honoureth me with their lips; but their heart is far from me.*

Isaiah 29:13 *Therefore the Lord said: "These people draw near to Me with their mouths and honor Me with their lips, but their hearts are far from Me. Their worship of Me is but rules taught by men.*

Acts 8:21 Thou hast neither part nor lot in this matter: for thy heart is not right in the sight of God.

Ephesians 4:11-14 And he gave some, apostles; and some, prophets; and some, evangelists; and some, pastors and teachers; For the perfecting of the saints, for the work of the ministry, for the edifying of the body of Christ: Till we all come in the unity of the faith, and of the knowledge of the Son of God, unto a perfect man, unto the measure of the stature of the fulness of Christ: That we henceforth be no more children, tossed to and fro, and carried about with every wind of doctrine, by the sleight of men, and cunning craftiness, whereby they lie in wait to deceive; But speaking the truth in love, may grow up into him in all things, which is the head, even Christ.

Satanism, Idolatry, & Atheism – From my manual *"Unmasking The Power Of The Scouts Manual II: Soul Stealers."*

Satanism entails a group of religious ideologies and philosophical beliefs based on Satan. These practices are designed to worship, serve, live in, and build the kingdom of Satan. Satan is a fallen angel that God kicked out of heaven because he pridefully exalted himself above God, and wanted to be God (*Ezekiel 28:1-26, Isaiah 14:2, John 12:31*). Satan has a host of other demons, fallen angels working with him, and people who revere his adverse, evil and wicked plans (*Jude 1:6, Revelation 12:1-17*). The Hebrew word for Satan is, "adversary." Satan and his disciples are against God, his people, and his principles. Once kicked out of heaven, he roams the earth seeking to kill, destroy, and devour the plans and purposes of God and God's people.

Job 1:6-7 Now there was a day when the sons of God came to present themselves before the LORD, and Satan came also among them. And the LORD said unto Satan, Whence comest thou? Then Satan answered the LORD, and said, From going to and fro in the earth, and from walking up and down in it.

John 10:10 The thief does not come except to steal, and to kill, and to destroy. I have come that they may have life, and that they may have it more abundantly.

John 8:44 You are of your father the devil, and your will is to do your father's desires. He was a murderer from the beginning, and has nothing to do with the truth, because there is no truth in him. When he lies, he speaks out of his own character, for he is a liar and the father of lies.

2Corinthians 4:4 In whom the god of this world hath blinded the minds of them which believe not, lest the light of the glorious gospel of Christ, who is the image of God, should shine unto them.

Occultism - From my manual "*Unmasking The Power Of The Scouts Manual II: Soul Stealers.*"

Occultism involves different ideologies and practices surrounding beliefs, knowledge, and the use of supernatural forces or beings, magic, divination, witchcraft, and alchemy. Occult practices include the ability to manipulate natural laws with the intent to navigate life experiences, produce luck, favor, fame, fortune, impact someone else's life, engage in wicked or evil practices. People form cults on all levels of society. On an extreme scale there are witchcraft occults that teach their practices, train people to be witches and warlocks, engage in witchcraft rituals, prayers, and operations. These groups may operate in secret creeds and codes that make it difficult for people to leave for fear of retribution. Retribution can be fear of being hurt due to the violence and spells the cult participates in, as many cult leaders threaten the lives of their members. There is also the impression that spell work will be done towards the person to make their lives miserable and full of demonic opposition. People feel a sense of loss of self, innocence, trust, destiny, and community and fear of integrating back into mainstream society, family, and relational structures. They experience guilt and shame regarding the belief systems and acts they have committed. Often people must be deprogrammed due to the dependence and inordinate dedication they have been groomed to have to the cult. They may experience post-traumatic stress disorder and other mental anxieties and disorders due to the trauma of being in the cult and leaving the cult. Intense counseling is needed to help such people reintegrate into normal life and society.

Children and teenagers learn how to form pacts and cults via movies, cartoons, TV shows, media, music, videos, video games, books, websites, social media, online groups, influence of their peers, and even parents and loved ones. They will reenact incantations, rituals, and spells, that they learned from these platforms. They will cut one another and make blood covenants that can include sexual experiences, pregnancy, suicide, and death. They will engage in self-injurious cutting and offer sacrifices to demons, idol gods, goddesses, the universe, and even their favorite Hollywood idols. Some children and teens have gone as far as being violent towards one another and have even succumb to murdering others. They are often unsupervised in these practices, so it makes it easier for demonic forces to influence and overtake them. The lack of biblical principles and practices governing the home, leave children and teens vulnerable to exploring alternative pathways to understanding their existence and connecting to a higher power. As the New Age movement grows, the body of Christ will need to educate themselves on how occults operate and how to deliver and heal those involved in these practices.

Shrines - Shrines are manmade. Most of them have idolatrous and pagan ties. Shrines are where people come to worship, fellowship, gain knowledge, train and equip, engage in altar sacrifices or witchcraft acts to enact sacrifices. A shrine can also be a place of honor and reverence to a person, someone famous, or deity. Demons, particularly familiar spirits will hang around shrines to lend honor for the purposes of attaching themselves to people who operate in idolatrous behaviors in relations to that person, famous idol, or deity.

Dictionary.com defines *shrine* as:
1. a building or other shelter, often of a stately or sumptuous character, enclosing the remains or relics of a saint or other holy person and forming an object of religious veneration and pilgrimage
2. any place or object hallowed by its history or associations: a historic shrine
3. any structure or place consecrated or devoted to some saint, holy person, or deity, as an altar, chapel, church, or temple
4. a receptacle for sacred relics; a reliquary

Covens & Sects – Covens are an assembly of witches. It is a where they gather to worship, offer sacrifices, conduct rituals, spells, create charms and potions, teach, train, etc. A sect is a body of people adhering to a particular religious faith or idolatrous doctrine. Some people will need to break soul ties with covens and sects, their leaders, vows, covenants, doctrines, and any curses that have been spoken on people who leave. Many covens have demons and physical people who track and terrorize those that leave. They will be tracked in their daily lives, spiritual realms, and dreams, while terrible death threatening experiences happen to them. They will be terrorized and even consider returning because they feel helpless in being free from them.

Paganism – Paganism is birthed out of what people deem is right in their own eyes. There is no regard for what God has said or established. The heart of man seeks to override the will and purpose of God, while deeming their ways essential in fulfilling their needs and the needs of others. This error even causes people to create a counterfeit version of what God has said and established.

Dictionary.com defines paganism as:
1. (no longer in technical use) one of a people or community observing a polytheistic religion, as the ancient Romans and Greeks
2. a member of a religious, spiritual, or cultural community based on the worship of nature or the earth; a neopagan
3. *Disparaging and Offensive*
    1. (in historical contexts) a person who is not a Christian, Jew, or Muslim; a heathen
    2. an irreligious or hedonistic person
    3. an uncivilized or unenlightened person

Merriman-Webster.com states the following:

Pagan is derived from the Late Latin paganus, which was used at the end of the Roman Empire to name those who practiced a religion other than Christianity, Judaism, or Islam. Early Christians often used the term to refer to non-Christians who worshiped multiple deities. Paganism entails worshipping multiple gods which is idolatrous and usually includes perverse, twisted, inordinate, ungodly witchcraft practices, traditional events, and rituals. There can also be concept of reverencing nature, the universe, and a revival of and animistic religious practices which is the belief that natural objects, natural phenomena, and the universe itself possess souls.

Pagans like to create traditional events, festivals, and holidays to ensure the gods they worship are receiving the offerings, sacrifices, and reverence they deserve. Pagans have a lot of false ideologies regarding how their gods will or will not respond based on their dedication and timeliness to serving and worshipping them. Many of these concepts are rooted in superstitions, fears, entertaining demons, religious mixture, and ancestral worship. Many pagan practices and events are tied to self-seeking pleasures, gains, financial advancements as people give their wealth and material goods for self-glory, gratification, success, monetary, and life advancement. Such pagan activities, holidays, festivals, and events appear harmless but are so strongly rooted in idolatry, witchcraft covenants and dedications, demonization, that it is difficult to separate the foundational intent and purposes and the influence it is designed to have on those who participate in them. Many are so soul tied to paganism that they become angry, defensive, and offended when questioned about their actions and told that there is no way around paganism being ungodly. Many will use scripture to justify such acts, claim they are participating in the event but are doing it unto God, or they become straight rebellious in stating that they "are doing them," while still claiming it has no bearing on their spiritual walk or relationship with God. Many people steeped in witchcraft will have to renounce, repent, and fall out of agreement with paganism to experience complete deliverance. Deliverance from paganism will have to be practiced as a lifestyle even as it is dealt with in a deliverance session.

Witchcraft Practices – Witchcraft entails the practice of magic, especially black magic. It is the utilization of spells and the invocation of demons to bind people, families, ministries, businesses, organizations, land, atmospheres, climates, regions, and nations. Some people engage in witchcraft for entertainment, curiosity, or due to ignorance. Those that dedicate their lives to it use it to acquire personal success and advancement, power, fame, rank in spiritual realms, spheres, and to obtain high-ranking positions and platforms in the natural. For a more defined encyclopedia of Witchcraft practices, I recommend my manual, *"Unmasking The Power Of The Scouts Volume II Soul Stealers."*

Witch – From my manual *"Unmasking The Power Of The Scouts Manual II: Soul Stealers."*

A female who engages in magical powers, especially that of divination, sorcery, dark magic, white magic, spells, hexes, vexes, incantations, curses, chants, demonic vows, dedications, and ceremonies. Witches offer up these practices to idol gods in exchange for demonic assistance to aide in helping them to control, manipulate, intimidate, seduce, destroy people, places, and things. Witches use witchcraft to violate freewill. This is the reason being a witch and engaging in witchcraft is ungodly and unbiblical. They are seeking power, success, fame, platform, position, provision, wealth, healing, safety, protection, relationship, judgement, justice, and intervention. These desires are self-absorbed and self-focused and are done with the assistance of witchcraft manipulations and demons. There is no such thing as a good witch. God does not honor or regard witchcraft. God is an enemy of witchcraft. He is not telling people to engage in witchcraft, neither is he promoting or using witchcraft to fulfill his purpose in the earth. God put witchcraft to death from the beginning of time. His exact words were,
Exodus 22:18, *"Thou shalt not suffer a witch to live."*

God believed the best way to rid the earth of witchcraft was to kill the witch. He has a no compromise stance regarding witchcraft. This has never changed to this day. His judgment remains the same. Suffer Not Says the Lord. The word suffer means to tolerate. Though we are not physically going to go out and kill anyone, we do have authority in the spirit realm to not tolerate witches and witchcraft, and to release God's judgment on those who engage in witchcraft or operate as witches. When witches are casting spells and evoking evil prayers against God and his people, they are seeking to make sure we experience pain, hardship, and destruction. Many of them are not repentant or remorseful for how their actions impact anyone. They are seeking to fulfill whatever purpose they are aspiring at that time. A plan that is at the expense of others, especially God's kingdom. They are workers of the devil and no toleration for anything or anyone outside themselves. God, therefore, has no toleration for them (Study ***Isaiah 57:1-12***)

Micah 5:12 New International Bible *I will destroy your witchcraft, and you will no longer cast spells.*

Witchcraft can also enter people's lives through rebellion. This means anyone can SHIFT into operating in witchcraft at any time, including church folks. God's people – YES, CHRISTIANS - can engage in witchcraft.

1Samuel 15:23-24 *For rebellion is as the sin of witchcraft, and stubbornness is as iniquity and idolatry. Because thou hast rejected the word of the Lord, he hath also rejected thee from being king.*

<u>*Rebellion* is *qesem* in the Hebrew and means:</u>
1. divination (including its fee), oracle

2. (reward of) divination, divine sentence, witchcraft
3. witchcraft of the nations, Balaam of false prophets

When rebellion operates in members and ministries, it opens the door to witchcraft. Rebellious believers,

- Are the people's choice and choose the people's choice. They want leaders who will cater to their fleshly and sinful desires, who they can control and manipulate for their own purposes.
- Hear the word, but implement it how they see fit, according to their own desires, what benefits them and those they are aiming to please.
- Only please or serve God in measure. They make excuses for the reason they are not fully committed or obedient to what God purposes them to do.
- Train just enough to get a position or present as godly, but not enough to sustain in their daily living, destiny, or calling.
- Are interested in position, leadership, rulership, but not submission or honor to God or leadership.
- Implement witchcraft tactics to continue advancing in what they have built in their own strength and will, and to appease those who are flattered by their witchery.
- Operate in mixture, but claim their sinful, idolatrous, and witchcraft practices are of God.
- Utilize witchcraft practices in effort to foretell the future or to hear from God once he stops speaking to them.
- Tend to act first, then repent, pray, and seek to hear God after they realized they have disobeyed him, and are a risk to the consequences of their actions.
- Seek to manipulate and kill God's chosen leaders due to jealousy and fear that they will be replaced.
- Will continue in their position unrepentant to save face with the people while knowing God has rejected them.

Warlock - From my manual "*Unmasking The Power Of The Scouts Manual II: Soul Stealers.*"

Warlock is the name for a male witch. The definition of the word warlock is "*path breaker, deceiver, traitor, liar.*" These definitions derived from historical acts of foundational warlocks making a pact with the devil, while separating from God and religious practices. I would contend that warlocks are lockers. They hook and lock their prey down and hold them in bondage with their clever powers. These hooking locks have to break to free the prey of the warlocks bewitching. Warlocks are very militant and learn and operate in their craft in a very military type of system. The way they learn their trade and operate in their craft requires them to become a master and obtain certain ranks until they reach a level of mastery.

Warlocks craft in the dark arts (manic and spell work used to harm, control, or kill their victim), magic, sorcery, conjuring and necromancy. They are generally called necromancers, pit servers, high priests, witch doctors, shaman, and witches. Extreme warlocks operate in high level witchcraft that include altar worship and rituals, ceremonies, cosmic energy practices, spell casting, astral projection and communing with demonic forces and beings in realms and spheres, communing and covenanting with Satan, animal and human blood sacrificing, and sexual and sadistic rituals. The average warlock is overly self-absorbed, egotistical, and some are narcissistic. They pride themselves on being supreme and individualistic in their ideologies and behaviors, and many view those around them as subservient and opportunities to work their craft for the purpose of increasing their warlock rank. Some warlocks are so possessed by demons, they may manifest demonic traits, exhibit inhuman strength, and supernatural behaviors. High level warlocks and witches can operate as shape shifters where they can change their physical form at will or adapt behaviorally to any situation. Consider purchasing my manual, *"Crushing Warlock Opposition."*

Wizards – From my manual *"Unmasking The Power Of The Scouts Manual II: Soul Stealers."*

A wizard tends to be a male, who operates in skillful magic, mysticism, magical and mystical practices, or wizardry. Wizards tend to work as illusionists or in positions that allow them to perform acts of trickery and deception on people. Some of them make pacts with demons to perform their wizardry as they aim to be the best at their craft and engage in illusions that present them as masters of their craft. They rarely share their techniques with others because the more individualistic they are, the greater the fame, success, and fortune of their craft.

Spirit of Witchcraft & Demonic Powers – A spirit of witchcraft is a demonic strongman that works with those who operate in witchcraft to aid these practitioners in completing their demonic workings. Demonic powers are the powers from the spirit of witchcraft, from the demonic altars, and from the demonic idols and elements to which they are deriving their powers from. Elements are the universe, moon, stars, sun, earth, rocks, waters, they are praying to receive their demonic powers. These elements have divine energies that they draw from. As these practitioners draw strength and power from these elements, there is an alliance that forms between them, the elements, and the witchcraft demons. As practitioners use ley lines to astral project or to send demons or witchcraft to people, places, and things, they are able to control, intimidate, manipulate, seduce, inflict, steal, kill, and destroy whatever and whomever they are attacking as they release and conduct their assignments.

Witches, warlocks, and witchcraft practitioners are also able to astral project and operate illegally in the spirit realm because their soul is tied by chords on the astral ley line planes. They are tied to the elements and the power it gives them to fly and travel to do their demonic workings.

When people go to witchcraft practitioners and pay for workings to be released against people, places, or things for certain purposes, a witchcraft demon is deployed to inflict these workings on that person, place, or thing. These workings can be in the form of curses, dedications, covenants, as people will dedicate other people, even their family members. They will make covenants with these witches and warlocks or with idol gods so they can have their workings released against other people, places, or things. They have no regard for the harm and danger they are releasing and how it can impact generations.

The witchcraft demons that are released to deploy these workings are tied to the witchcraft altar. The witchcraft demon and sometimes the witch or warlock will astral project to that person, place, or thing, to attack and release their workings. They will attack a person in their dreams, while the person is sleeping, physically hit, and scratch the person, plant stuff in and around the home, etc. Sometimes the person is fighting the demon, witch, or warlock and wake up with bruises, scratches, soreness, pain, and afflictions. Demons and witchcraft practitioners have been known to try to brand or place a mark upon a person to identify them as belonging to them and their system. They may also use blood to write things upon you or mark you. This is a counterfeit act of how God use the sacrificed animals' blood upon the doorpost in Egypt so that the death angel would Passover those that were marked as God's people.

Exodus 12:23 For the LORD will pass through to smite the Egyptians; and when he seeth the blood upon the lintel, and on the two side posts, the LORD will pass over the door, and will not suffer the destroyer to come in unto your houses to smite you.

The devil wants to mark and claim us for his own. He also will retaliate in anger and strive to brand us because we reject him and his demonic kingdom.

Demons and their demonic practitioners love to try and feed people witchcraft and death in their dreams and sleep realms.

Mark 16:18 They shall take up serpents; and if they drink any deadly thing, it shall not hurt them; they shall lay hands on the sick, and they shall recover.

The Greek definition for *deadly* is "*poisonious or fatal.*" The devil wants people to eat and drink things that will kill them. He tricked Eve and Adam into eating from the tree of knowledge of good and evil. He tried to get Jesus to eat bread during a 40 day fast. Jacob sold his birthright for food. Demons and witchcraft practitioners know that people love the pleasure of eating. They will enter their dream and sleep realm and feed people witchcraft and release demonic deposits into the person. They want to feast upon the flesh of people through these attacks and cause the flesh to be eaten up so the person can perish. These demon deposits that are attached to demons and witchcraft practitioners can be in the form of witchcraft spells, charms, potions, snakes, worms, parasites, demons, poison, sicknesses, afflictions, diseases, fear, confusion, madness, mental illness, emotional instability, sabotage, tragedy, failure, imprisonment, poverty, loss, death, etc. God will deal with these demonic feeders.

Psalm 27:1-2 The LORD is my light and my salvation; whom shall I fear? the LORD is the strength of my life; of whom shall I be afraid? When the wicked, even mine enemies and my foes, came upon me to eat up my flesh, they stumbled and fell.

Sometimes what is being imparted will manifest as a dream so the person will think they are experiencing a dream and not a real impartation. This is the reason after such dreams, demonic and challenging experiences occur. People are experiencing the physical manifestation of what was imparted in the dream and sleep realm. These demons and wicked people will sleep with the person and release demonic sperm, seed, eggs, or substances, while masturbating, molesting, having sex with, or raping the person. They will physically hit and scratch the person and use the skin, blood, hair, particles, body fluids, from the person upon witchcraft altars for further witchcraft hexing. They will steal pictures, jewelry, and other items to use upon the witchcraft altars.

These types of attacks have increased as I am having more and more people come to me for breakthrough in this area. They are not accustomed to how the spiritual realm or witchcraft works so it will take time to help them understand what is happening to them, pull down the strongholds that has them bound in disbelief and a lack of faith, and deal with the strongman behind their attack. Many of them dread the warfare and work it takes to be free. It will take time to counsel them into having a true relationship with the Lord where they love not their life but desire the life he has for them. Consistently batter the following until the person is free.

- Break contracts, curses, dedications, and covenants that have been spoken over or about your life, destiny, calling, generational line, land, region, and/or sphere of influence.
- Command everything the demon, warlock, and/or witch has been attacking the person with, put in the person, or on the person to be removed. Command all of it to be removed from the person's life. The deliverance minister may have to be specific in commanding sicknesses, substances, sperm, charms, snakes, fear, etc. to leave the person and even commanding it out of different parts of the person's mind, heart, soul, body, dream realm, sleep realm, situations, etc.
- Command the spirit of witchcraft to leave, witchcraft powers to break, and any demons working with it to be bound and cast out.
- Command every other strongman and the demons working with them to be bound and cast out. Deal with them as God reveals them to you and command them to go.
- Break the powers of witchcraft from the person and break every way they are tied to the altar.

- Command the soul ties and powers of the person who requested the witchcraft and that of the witchcraft worker to loose the person.
- Break their powers and silver cord from the elements, astral planes, and ley lines and command their strength and ability to astral project and attack to be dismantled, disbanded, and cut off in Jesus' name.
- Command everything that is on altars anywhere concerning the person to be burned up and destroyed.

Witchraft From Believers – Many people experience witchcraft because there are people who claim to be God's people – believers – praying for horrible things to happen to other believers. They will claim this is God's judgment, but really it is their own flesh, unstable mind, or demonic oppression at work. They become witches and warlocks who release fiery darts of death and evil against other believers while claiming to be of God and to be working for God. These people will become so overtaken by the demonic oppressions they are under that they will begin to use the bible and God's principles like witchcraft, while claiming their actions are godly. They will use them like spells, hexes, vexes, incantations, and root workings. They will engage in numerology and claim the numbers are aligning them to bible scriptures that confirm what they are saying. They are so deceived that if you try to correct them, they will contend you're being disobedient and defiant towards them and their calling, and will pray and release further demonic words, curses, and judgments against you. They will hook weak, ignorant, fearful, immature people into their web. These people who are afraid of their operations and/or cannot rightly divide the word of truth, will be coerced to side with them and have them praying, cursing, and judging you. As these people form an alliance, they will not adhere to those who will try to bring correction to them. They will confirm one another's witchcraft workings, while rejecting and bewitching anyone who is resistant to their evil workings. This will increase in the earth as free speech becomes more and more prevalent and people do not need to be discipled, taught, equipped, or held accountable for their actions. They will be able to speak and preach whatever they want with no regard for how inaccurate they are, evil they are, and far from God they are. The Bible lets us know that this will occur. It also lets us know that many believers will be led astray by these so call believers who are really witchcraft practitioners.

Matthew 24:24 For false messiahs and false prophets will appear and perform great signs and wonders to deceive, if possible, even the elect.

2Timothy 3-4 For the time will come when they will not endure sound doctrine; but after their own lusts shall they heap to themselves teachers, having itching ears; And they shall turn away their ears from the truth, and shall be turned unto fables.

1John 2:26 These things have I written unto you concerning them that seduce you.

1John 4:1 Beloved, believe not every spirit, but try the spirits whether they are of God: because many false prophets are gone out into the world.

2Peter 2:1 But there were false prophets also among the people, even as there shall be false teachers among you, who privily shall bring in damnable heresies, even denying the Lord that bought them, and bring upon themselves swift destruction.

Bewitchment – Bewitchment means to cast a spell, hex, vex, on someone.

Galatians 3:1 O foolish Galatians, who hath bewitched you, that ye should not obey the truth, before whose eyes Jesus Christ hath been evidently set forth, crucified among you?

The Message Version You crazy Galatians! Did someone put a hex on you? Have you taken leave of your senses? Something crazy has happened, for it's obvious that you no longer have the crucified Jesus in clear focus in your lives. His sacrifice on the Cross was certainly set before you clearly enough.

Bewitch in *baskaino* in Greek and mean:
1. to malign, i. e. (by extension) to fascinate (by false representations)
2. to bewitch, to speak ill of one, to slander
3. traduce him to bring evil on one by feigning praise or an evil eye to charm

Traduce means to speak maliciously and falsely of, to slander, defame, to traduce someone's character.

Bewitchment binds people where they cannot obey the truth. They become contrary to him, his word, his standards, and his kingdom, causing them to misalign with God. Some people who are bewitched experience headaches, doublemindedness, mental and psychological confusion, fear, anxiety, demonic dreams, demonic encounters, misinterpretation, and error when appropriating God's word, spiritual gifts, and callings. The bewitchment will have to be broken so they can return to operating through the mind of Christ and the truth of God.

Ley Lines - Ley lines are demonic supernatural traffic and workings. Ley lines are known as magical energy lines in and around atmospheres. Ley lines are supernatural highways aligned in spiritual realms and spheres. They are positioned according to planets, the elements (stars, moon, sun, earth, universe), and specific designated access points, monuments, sacred sites, within regions and spheres that have been dedicated to idols. Ley lines are known to be full of strong electromagnetic fields – supernatural energy - that can influence mood, perception, memory, and behavior.

People who engage in witchcraft, the new age movement, operate through a higher consciousness, believe in god the universe, etc., utilize and draw energy – supernatural demonic knowledge and powers from upon ley lines and the demons that control ley lines.

Witches, warlocks, high priests, clairvoyance, astral projectors - out of body experiences, demonic visions, and precognitions, use ley lines to astral project their soul into spiritual realms and travel to wherever they want to go to do their witchcraft workings and dealings. They can open third eyes (also known as the sixth chakra or anja), portals, and demonic gateways, send demonic dreams, witchcraft spells, hexes, vexes, hoodoo, incantations, curses, demonic prayers and chants, root workings, etc. to people, places, and things using these ley lines. This is how you get hit with demonic attacks and energies, sexual energies, witchcraft dealings, that are being sent to bind, blind, oppress, stronghold, and soul tie you.

When you are in public places and are trying to figure out why needle pricking's, sicknesses, headaches, slime and cobweb sensations are hitting you – it is or has been sent to that land, atmosphere, or to you by way of the ley lines. The land and atmosphere could also be dedicated, and you are picking up on the idolatry or witchcraft that resides there. The territorial spirits, watcher or squatter spirits on the land or witches and warlocks tied to the land could have spotted you in the spirit and are attacking you with witchcraft because they deem you a threat.

This is how confederate attacks operate against you as you travel from region to region and even just throughout your life. They track your habits, behaviors, whereabouts, etc., and send intel upon the ley lines regarding how you operate and move about. They send demons to ambush you so you cannot be successful or have warfare and challenges in whatever you are doing. Or they notify territorial demons in that area and region so they can be on high alert regarding you; so they can frustrate, hinder, weary, delay, or stifle your reasoning for coming to that area.

When you are hearing demonic chatter, experiencing psychological, mental, and physical warfare, it is coming by way of let lines and through the frequencies and airways in and around you.

Words, curses, prophecies, chants, prayers, spells, basically anything spoken has the ability to live upon ley lines and within frequencies and airways. This is the reason the Bible tells us to guard our words – to be careful what we speak. What we speak has a life and supernatural power of its own. What we speak is meant to breathe life, to produce life, to manifest as substance. God demonstrates this in Genesis when he spoke, and his presence moved upon and formed the earth. What we speak can become a vow, law, decree, a manifestation, or simply linger in spiritual realms. These words are not nullified and cleansed out of these areas until we cancel them.

Sidebar Revelation: The pineal gland is often referred to as the "third eye" and is also referred to as the mystical gland. The pineal gland is in the deep center of the brain. It helps the body make melatonin and is key to helping the body control sleep patterns. People actually spend time fasting and cleansing their pineal gland so they can have keen demonic eyesight and discernment. This is the reason psychics, high priests, witches, and warlocks, can speak accurately regarding your life. They purify themselves and do what they call, emptying their minds and consciousness through meditations so their senses can be keen, also so high-ranking demons can inhabit them and operate through their third eye or through their seer gift that should be used for God.

People operate in a false sense of security through the false peace, contentment, safety, security, success, that is being set up on the ley lines by demons and demonic systems they have yoked themselves to. They have no clue that an entire system is operating to give them a false sense of hope from what they or their generational lines have yoked themselves to, so they can serve and covenant with idol gods.

Nehemiah 4:7-9 But it came to pass, that when Sanballat, and Tobiah, and the Arabians, and the Ammonites, and the Ashdodites, heard that the walls of Jerusalem were made up, and that the breaches began to be stopped, then they were very wroth, And conspired all of them together to come and to fight against Jerusalem, and to hinder it. Nevertheless we made our prayer unto our God, and set a watch against them day and night, because of them.

This is the reason we must be Gatekeeping the walls of prayer. So, we can be proactive in stopping the acts conspiring in spiritual realms against us.

It is essential to pray to shut down ley lines and cut and close off their access points. Nullify anyway these ley lines have connected and soul tied to you where they have access to your home, bloodline, dream and sleep realm, soul, mind, body, especially your consciousness, eyes, ears, senses, pineal gland, central nervous system, spine, sexual organs, stomach etc. They target these areas so they can oppress, stronghold, harass, and feed you their witchcraft.

This is the reason the devil goes so hard after prophets and seers. He wants your body and giftings. He wants to house and influence you so he can use your gift for his kingdom.

1Corinthians 14:1 tells us to, "*Follow after charity, and desire spiritual gifts, but rather that ye may prophesy.*"

The Amplified Bible EAGERLY PURSUE and seek to acquire [this] love [make it your aim, your great quest]; and earnestly desire and cultivate the spiritual endowments (gifts), especially that you may prophesy (interpret the divine will and purpose in inspired preaching and teaching).

The devil knows the power and authority of having insight, hindsight, foresight, and knowing the divine will and purpose of a person. Whoever has this power and authority can control and govern over their success and the success of others. While we eagerly desire not to prophecy and see, he eagerly desires to possess us so he can use our gift to his advantage.

<u>**Portals**</u> – A portal is a door, gate, or entryway that can be established in a natural or spiritual realm.

In the natural realm, witches, warlocks, demons, mark where they desire portals to be. They dedicate areas by using witchcraft works. Often altars are erected in these areas. Demonic prayers, chants, and sacrifices are made consistently to keep the witchcraft workings active. Sometimes demons are summoned to these areas to guard, protect, oppress, bind, that area and the people who enter the portal. These portals can operate as crossroads such that when people enter or cross the witchcraft line, it activates different witchcraft workings and activities. Depending on what witchcraft workings were released, people can be bewitched, enticed to do various things, and even sacrificed unto death to idol gods.

In spiritual realms, portals include different worlds (regions and spheres) people, angels, demons, supernatural beings can enter. They also can consist of planets.

Portals can also be opened in people's:

- *Mind* (In their consciousness, memory recall, amygdala, pineal gland, skull); People will hide inside the spiritual and fantasy realms in these regions and spheres of their brain. They will go in and out of these realms to dissociate from trauma experiences, responsibilities, accountabilities. Some people can get stuck inside these realms and spheres which can cause alter egos, split personalities, autism, etc.

- *Dream and sleep realm* (Hippocampus which is the region of the brain that causes sleep and dreams; Pineal gland which is in the deep center of the brain. It helps the body make melatonin and is key to helping the body control sleep patterns). Demons, witches, and warlocks use these portals to enter people's spiritual realms via astral projection to attack people as they sleep. Or they use the portals to manipulate or enter their dream realms and cause demonic, horrible, sexual, or challenging dreams. Witchcraft spells, demonic assignments, energies, and vibes can be sent through the dream and sleep realm portals as well. This can cause people to think upon others at night, cause them to engage in sexual acts or masturbation in their sleep, sleepwalk, trances, sleep paralysis, stop breathing in their sleep (sometimes but not always the case, sleep apnea is caused by this).

- *Heart & Soul* (These portals are open through issues of the heart and wounds in the soul. Demons, witches, warlocks, use these openings to lay claim to people. They use these portals to sustain their soul ties, and release demonic energies and witchcraft workings, thus keeping people in bondage to them. If a person is soul tied to another individual, these portals can be used to unknowingly or knowingly keep that person tied to the other person. Transferences of demonic activities, witchcraft practices, sin issues, energies and vibes, substances, ungodly cycles and patterns can transfer through the soul tie depending on the interactions that occurred in the relationship. This could be a factor in how someone has a difficult time letting go of a person who is no longer in their life. This portal needs to be closed and the issues listed above need to be dealt with to break the holds off the person and SHIFT them into deliverance and healing.

Third Eye – From my manual Soul Stealers: The third eye illegally operates through the pineal gland that is located in the deep center of the brain. The third eye is the seer operation of opening a gateway or portal within the eye gates, imagination, psyche, and inner realms of the second heaven. The person opens these pathways to receive information illegally through demonic assistance, for the purposes of engaging in witchcraft operations, and to ascend to higher states of consciousness. The third eye is also called the mind's eye or inner eye and is often depicted as being on the forehead between a person's physical eyes. It is mysterious, demonic, and abstruse, in nature because it is an invisible eye that is used for evil purposes, or for perceived good purposes that is rooted in demonic operations. People receive mental images, visions, revelation, and guidance that they believe has some deep spiritual significance, but these insights derive from demonic spirits that have tracked their lives, or that are using familiarity to transmit information, enlightenments, vibes, energies, powers, and sensations from which the person desires to experience.

When Satan dealt demons to Eve in the garden, he opened her to a third eye (*Genesis 3:4-7*). Satan told Eve her eyes would be open to know good and evil if she partook of the tree. When doing so, it opened a third eye where she begin to see and experience herself, her husband, and the world in ways that was not of God. It was forbidden and not design or purpose of God. Third eyes are presented as needful, godly, and essential, but are opened through demonic means that are contrary to God.

Dream & Night Season Trauma – Some people will pursue deliverance due to demonic or ungodly attacks during the night and in their dreams. Many people assume that these attacks are due to incubus and succubus spirits or witchcraft.

- **Incubus** is a male demon that lies upon sleeping women and attack them sexually.
- **Succubus** is a female demon that lies upon a sleeping man and attack them sexually.

  Exodus 22;16 And if a man entice a maid that is not betrothed, and lie with her, he shall surely endow her to be his wife. If her father utterly refuse to give her unto him, he shall pay money according to the dowry of virgins. Thou shalt not suffer a witch to live.

  Incubus and Sucubus demons lay claim to you via sexual acts with people willingly or unwillingly as fornication, adultery, oral sex, masturbation, rape, incest, molestation, sexting, pornography, sadomasochism and other perverse acts, secular music, ungodly movies, satanic and Greek organization, etc.

Dream and night season attacks can also occur via astral projectors who engage in witchcraft practices or thrill seek in supernatural realms for various ungodly purposes. They will have sex with people, engage in tricks and ghostly behaviors. See the area on astral projection for more detailed information in this area.

- **Witches, warlocks, rulers of darkness**, and ignorant people who engage in witchcraft practices release a lot of witchcraft in the night season. These practices can be released to target specific people, or released in general against certain types of people, groups of people, particular communities, and regions. Sometimes witchcraft practices are done as sacrifices to idol gods and serve as a gateway to demons releasing their assignments in the earth. Such practices can linger on frequencies, airways, atmospheres, lands, and then manifest upon people due to open doors or afflictions upon the righteous.

  Proverbs 6:22 *As the bird by wandering, as the swallow by flying, so the curse causeless shall not come.*

Though this may be the case, it is important to seek God for revelation on the root issues of these attacks as there could be other factors at work, such as:

- Different demonic spirits (predator spirits, familiar spirits, destiny killing spirits, spirits of fear, terror, and torment).

- Astral projectors are people who knowingly or unknowingly use their soul to leave their body and engage in spiritual activities. Sometimes they will visit people, places, and destinations for various reasons; have sexual encounters with people while they sleep; scare or haunt people and buildings for tricks and fun; spy or gain intel on people, lands, and regions; release spells and witchcraft workings, engage in worldly and demonic trading and business endeavors; receive education and training in demonic and worldly realms; visit demonic kingdoms, covens and altars to aid and assist with demonic and demonic kingdom operations, etc. These factors all depend on their purposes for astral projecting. New Agers engage in astral projection to "mind trip" which basically means for fun and thrills. They also astral project for the purposes of engaging in various witchcraft practices, engage demons for enlightenment or higher consciousness, to maneuver from one place to another, etc.

- Personal and generational curses, vows, altars, covenants, and dedications.

- Idolatry and witchcraft practices in the family line opens demonic portals and gateways for such attacks.

Dreams are real realms and what occurs can manifest in the natural. It is essential to cancel and nullify demonic and ungodly activity in dreams and deliver people from dream attacks and infestations.

Night Terrors – Are frightening fear demons and dreams that seek to instill fear, anxiety, stress, hopelessness, helplessness, mental illness, and trauma into people. ***Study Psalm 91***.

From my manual, "Let There Be Sight."

Proverbs 3:24-26 *When thou liest down, thou shalt not be afraid: yea, thou shalt lie down, and thy sleep shall be sweet. Be not afraid of sudden fear, neither of the desolation of the wicked, when it cometh. For the Lord shall be thy confidence, and shall keep thy foot from being taken.*

The word *fear* in this scripture is *"pahad"* in the Hebrew and means, *"sudden alarm, terror, dread, great dread, an object or thing of dread."*

The Amplified Bible *When you lie down, you shall not be afraid; yes, you shall lie down, and your sleep shall be sweet. Be not afraid of sudden terror and panic, nor of the stormy blast or the storm and ruin of the wicked when it comes [for you will be guiltless], for the Lord shall be your confidence, firm and strong, and shall keep your foot from being caught [in a trap or some hidden danger].*

In this passage of scripture God promises that if we are not afraid when we lie down, our sleep will be sweet. He encourages us to not be afraid (startled, alarmed, caught off guard) of sudden terror and panic, and that He will even keep our foot firm and from being overtaken.

- These terrors can be a demon, sudden fear due to challenging situations, or just a panic of fear that hits us out of nowhere.
- These sudden terrors can come during the day and at night while we are sleeping.
- The Hebrew word for *foot* is *"regel"* and means *"to be able to endure, haunt, journey, keep pace."* These terrors come to haunt us, steal our journey, and prevent us from keeping pace in God. They will try to kill and overtake us and hinder, thwart, and scar us from walking in the calling that is on our lives or even just functioning daily in life.
- These terrors will make us feel as though we have done something wrong, but God tells us *"You will be guiltless."*

If you are not participating in any unrepentant sin, witchcraft and idolatry that may cause these attacks to occur, knowing you are guiltless is important when experiencing these attacks. I say this because the enemy and people will make you feel like you have done something wrong or opened a door, and that is the reason they are occurring. I have been accused of all kinds of things in my effort to acquire deliverance from demonic dreams, visions, and demonic night attacks and visitations. These accusations only caused me to be more fearful, more frustrated, and angry with people who should have been able to help me, and with God who I expected to deliver and protect me. I had to realize that these attacks and encounters were due to and a part of the calling on my life. When I received and accepted that revelation, I began to cleanse myself of the anger, condemnation, and shame and guilt that I carried. Then with the help of the Holy Spirit, I began to teach myself how to reign over my night and dream realms and be proactive in my attacks against the enemy.

> *Psalm 91:1-6* *He that dwelleth in the secret place of the most High shall abide under the shadow of the Almighty. I will say of the Lord, He is my refuge and my fortress: my God; in him will I trust. Surely he shall deliver thee from the snare of the fowler, and from the noisome pestilence. He shall cover thee with his feathers, and under his wings shalt thou trust: his truth shall be thy shield and buckler. Thou shalt not be afraid for the terror by night; nor for the arrow that flieth by day; Nor for the pestilence that walketh in darkness; nor for the destruction that wasteth at noonday.*

In *Psalm 91*, David is declaring that he dwells in the secret place (covering) and shadow (presence) of God. This secret place and shadow possesses:

- God's protective wings (feathers)
- God's covering (refuge)
- God's force field or hedge (fortress)
- God's revelation (shield)
- God's truth (buckler)

David is also declaring that because of God's covering and presence, he will be delivered from and will not fear:

- The snare of the fowler (traps)
- The noisome pestilence (aggravating and pesky demons and experiences)
- The pestilence that walks in darkness (sicknesses, afflictions, demons, ghosts, astral projectors which are people, witches, warlocks who translate themselves in the spirit realm to do harm or for thrills, evil lurking in the dark or behind the scenes of life)
- The terror by night (fear, dread, demons, and situations that haunt us)

- Arrows that fly during the day (situations, word curses, witchcraft, demonic attacks that hit out of nowhere and for no apparent reason)
- The destruction that waste at noon day (situations and demons that come to cause destruction)

Since David is in the secret place, what reason does he have to remind himself or make a declaration that God will surely deliver him and that he should have no fear? That's a great question. I was challenged by the realization of what David was experiencing despite being in the secret place (covering) and shadow (presence) of God.

Astral Projection - Astral projection is an out-of-body experience, during which one's soul separates from the physical body and traverses the astral plane with the intent of traveling to different places and spheres. An astral plane is a literal road type plane or world that the soul crosses over into for the purposes of astral projecting (traveling) to different areas in the spiritual realm. A silver cord is a life-giving cord that connects people's soul to their physical body. The person is separated from their body but is still connected by a silver cord similar to an umbilical cord that attaches a baby to its mother's placenta.

Ecclesiastes 12:5-7 Also when they shall be afraid of that which is high, and fears shall be in the way, and the almond tree shall flourish, and the grasshopper shall be a burden, and desire shall fail: because man goeth to his long home, and the mourners go about the streets: Or ever the silver cord be loosed, or the golden bowl be broken, or the pitcher be broken at the fountain, or the wheel broken at the cistern. Then shall the dust return to the earth as it was: and the spirit shall return unto God who gave it.

When a baby is born, the umbilical cord is cut from the placenta then a clamp is put on the remaining stump to pinch it off. In the womb the umbilical cord delivers oxygen and nutrients to the baby so it can grow. After birth, the clamped cord heals and the clamp falls off on its own, thus forming an umbilicus (belly button). The belly button is where the silver cord connects the person's soul to their body. If the silver cord is cut while the person is astral projected, they could die if they do not SHIFT back into their body quickly. This is a fear that witches, warlocks, and astral projectors have, and is a weapon that a person can use when encountering astral projectors who attack them while they sleep, in their dream realm, or in spiritual realms.

Astral projectors are people who knowingly or unknowingly use their soul to leave their body and engage in spiritual activities. Sometimes they will visit people, places, and destinations for various reasons; have sexual encounters with people while they sleep; scare or haunt people and buildings; for tricks and fun; spy or gain intel on people, lands, and regions; release spells and witchcraft workings, engage in worldly and demonic trading and business endeavors; receive education and training in demonic and worldly realms; visit demonic kingdoms, covens and altars; to aid and assist with demonic and demonic kingdom operations, etc. It just depends on their purposes for astral projecting. New Agers engage in astral projection to "mind trip" which basically means for fun and thrills, for the purposes of engaging in various witchcraft practices, to engage demons for enlightenment or higher consciousness, to maneuver from one place to another, etc.

Astral projection is different than translating via the Holy Spirit. In astral projection, the soul is illegally separating from the body by:

- Use of constant consciously focused meditation.
- Demonic assistance.
- Wooing or drawings into spiritual realms by demons and people they are soul tied to.
- Generational witchcraft practices, dedications, and covenants that have tied the souls of members of a family line to demons and altars, thus enabling their soul to astral project whether at will or for a specific purpose. Sometimes these people will find themselves hovering over their bodies, suspended in the clouds and heavenly spheres, particularly the second heaven, visiting covens, altars, demonic kingdoms, especially those under the waters, engaging in sexual acts in their dreams with people and/or demons or on altars while these acts are occurring, at dinner tables or altars while being served food as this is how demons, witches and warlocks keep them tied to these witchcraft practices, past lands, houses, and familiar areas that their ancestors frequented, especially those who engaged in sin, witchcraft, and idolatry.

Astral projection is illegal because there should never be a separation between our bodies, souls, and spirits. They should always remain one entity. Also, our bodies and souls should be subject to our spirit so that they can be guided and led by the Holy Spirit which is God's presence that lives on the inside of us. The Holy Spirit guides us into all truth and teaches us what is godly versus demonic/ungodly, legal versus illegal, healthy verses unhealthy, for us. Our soul also is to be knitted to God and restored in covenant relationship once we accept him as our personal savior. Our soul therefore should never be governed by anything or anyone outside of God, not even our own selves.

The Holy Spirit does not astral project us out onto astral planes, but does translate us into spiritual realms. Translate means *"to bear, carry, or move from one place, position, etc., to another; transfer."* When we are translated by the Holy Spirit, he lifts our entire being (body, soul, and spirit) to navigate us into and through spiritual realms. There is no separation of body, soul, and spirit, as this is a supernatural ascending into spiritual realms legally guided by the Holy Spirit. Astral projection is illegal because it is done by the soul, causes separation, and is guided by the person, their consciousness, or demonic spirits.

Sometimes, God will give people the authority to translate at will – as they feel led. Such a person can be entrusted to operate legally in spiritual realms regarding kingdom matters. They are mature enough to handle this level of favor and spiritual governance from the Lord. Many apostles, prophets, seers, and intercessors who scout out worldly and demonic realms for intel, may be given this liberty by God. But if God does not assign this right to a person, they should only engage in translation as the Holy Spirit leads.

The Holy Spirit tends to lead the general population of believers in translation during dreams, visions, and intercessory prayer. He will lift their spirit up or automatically translate them to a location for spiritual purposes or because he knows people will listen and pay attention more effectively if they are having a supernatural encounter. Sometimes people may not even know this is occurring or how they got to where they were in dreams, visions, and spiritual realms.

It is also important to note that as believers, we should be endeavoring to live, abide, and constantly ascend into heavenly places with Christ Jesus. The spiritual realm is eternal, and we are to have dominion in them even as we do on the earth. So, during prayer, we should be seeking to be lifted where we abide in our established seat of sonship above principalities and powers, live and operate from this place, and SHIFT from level to level and glory to glory via spiritual realms.

From my manual, "*Ascending Into Heavenly Realms.*"

> ***Ephesians 2:6*** *contends that we are seated in heavenly places in Christ Jesus. And hath raised us up together, and made us sit together in heavenly places in Christ Jesus. And hath raised us up together and made us sit together in heavenly places in Christ Jesus.*
>
> ***The Amplified Bible*** *And He raised us up together with Him and made us sit down together [giving us joint seating with Him] in the heavenly sphere [by virtue of our being] in Christ Jesus (the Messiah, the Anointed One).*
>
> Set is *sygkathizō* in Greek and means:
> 1. to give (or take) a seat in company with
> 2. (make) sit (down) together.
> 3. to cause to sit down together, place together
> 4. to sit down together
>
> Dictionary.com defines *set* as:
> 1. to put (something or someone) in a particular place
> 2. to place in a particular position or posture
> 3. to place in some relation to something or someone
> 4. to put into some condition

5. to put or apply
6. to put in the proper position
7. to put in the proper or desired order or condition for use
8. to post, station, or appoint for the purpose of performing some duty

Heavenly places are heavenly regions. They are the supernatural realms and dimensions that we refer to and must recondition our mind to live in and from.

Since this is the finished work after the cross, the word *set* denotes an eternal place - an eternal authority and governing in God - an established seat of authority inside the heavenly realms in Christ.

Through your established seat, you commune with Jesus Christ as he delivers, heals, and transforms you into Christlikeness.

2Corinthians 3:18 But we all, with open face beholding as in a glass the glory of the Lord, are changed into the same image from glory to glory, even as by the Spirit of the Lord.

New Living Bible So all of us who have had that veil removed can see and reflect the glory of the Lord. And the Lord—who is the Spirit—makes us more and more like him as we are changed into his glorious image.

As we mature in Christ likeness, Jesus Christ reveals the mysteries and realities of the kingdom of heaven to you, such that your seat expands and he allows you to be privy to heavenly visions, wisdom, revelation, knowledge, counsel, might, judgements, justice, translations, the literal kingdom of heaven, storehouses, weaponry, and provision.

We are eternal beings - we are not of the earthly world. From the beginning we were meant to live from a posture of eternal dominion.

Genesis 1:28 And God blessed them, and God said unto them, Be fruitful, and multiply, and replenish the earth, and subdue it: and have dominion over the fish of the sea, and over the fowl of the air, and over every living thing that moveth upon the earth.

Psalm 8:4-6 What is man, that thou art mindful of him? and the son of man, that thou visitest him? For thou hast made him a little lower than the angels, and hast crowned him with glory and honour. Thou madest him to have dominion over the works of thy hands; thou hast put all things under his feet.

Philippians 3:20 *But our citizenship is in heaven, and we eagerly await a Savior from there, the Lord Jesus Christ.*

James 4:4 *You adulterous people! Do you not know that friendship with the world is enmity with God? Therefore whoever wishes to be a friend of the world makes himself an enemy of God.*

Dominion denotes a kingdom throne of reigning, ruling, governing, and prevailing in the spirit and authority of God. Jesus came to restore the dominion that crumbled through the fall of man. Even as Jesus Christ walked the earth as a natural man, he made it clear that he was not of this world, and we were not of this world. He maintained a heavenly perspective and was a demonstrator of dominion - of how we should live daily from this truth.

John 17:14-19 *I have given them your word. And the world hates them because they do not belong to the world, just as I do not belong to the world. I'm not asking you to take them out of the world, but to keep them safe from the evil one. They do not belong to this world any more than I do. Make them holy by your truth; teach them your word, which is truth. Just as you sent me into the world, I am sending them into the world. And I give myself as a holy sacrifice for them so they can be made holy by your truth.*

Without a dominion posture and perspective, you will be inconsistent with living and governing from a heavenly perspective. You may have a mind to rule but will conduct your rulership from the earth and the second heaven.

Ephesians 1:15-23 *Wherefore I also, after I heard of your faith in the Lord Jesus, and love unto all the saints, Cease not to give thanks for you, making mention of you in my prayers; That the God of our Lord Jesus Christ, the Father of glory, may give unto you the spirit of wisdom and revelation in the knowledge of him: The eyes of your understanding being enlightened; that ye may know what is the hope of his calling, and what the riches of the glory of his inheritance in the saints, And what is the exceeding greatness of his power to us-ward who believe, according to the working of his mighty power, Which he wrought in Christ, when he raised him from the dead, and set him at his own right hand in the heavenly places, Far above all principality, and power, and might, and dominion, and every name that is named, not only in this world, but also in that which is to come: And hath put all things under his feet, and gave him to be the head over all things to the church, Which is his body, the fulness of him that filleth all in all.*

<u>Legal God ordained experiences of spiritual translations found in the Bible:</u>

- Prophet Ezekiel speaks of being taken up or snatched up out of his body by the spirit of God (*Ezekiel 3:14, Ezekiel 8:3, Ezekiel 37:1, Ezekiel 43:5*).
- Apostle Paul speaks of having out of body experiences where he was translated to heaven (*2Corinthians 12:2-4*).
- Prophet Elijah was taken up to heaven in a whirlwind and never returned (*2Kings 2:3-12*).
- Phillip the eunuch, was snatched away never to be seen again (*Acts 8:39*).

The experiences I listed above were not led by the soul but through the guided translation of the spirit of God – the Holy Spirit. *Study them and ask God to give you further insight and favor to translate and operate in spiritual realms.*

Sometimes a gateway to astral projection can be opened in the generational line which will cause a person to astral project without trying or even knowing it is occurring. This was my experience. I had to close these witchcraft gateways via generational repentance and intercession to stop my soul from astral projecting as I slept at night. Because this was a generational gateway, I would wake up and find myself suspended in spiritual realms. I would also encounter witches, warlocks, demons, and regular people on the astral planes. The demons and witchcraft practitioners could see the light of God around me and would often physically attack me or follow me back to my home and strangle me, hold me down, and fight me. They would also wait until I was asleep and/or return later and visit me, while attacking me in my dreams and sleep. They tracked me and was making sure to intimate and cower me because of my spiritual abilities to translate, see, and operate in God's power and authority against them. They would send spells, hexes, and vexes to my home, life, and literal body to oppress and afflict me. I became a huge target in the demonic realm and had not even been saved for real yet and had not begun my ministry.

When I did get saved for real and started conducting ministry, the attacks increased. I was already targeted as a threat and was put on the devil's hit list. I am still on that list to this day as I encounter spiritual warfare that would have killed the average believer. It took me years to understand what these attacks were about, the reason they were occurring, and how to stop them. Now that the illegal doors are closed, and though the enemy will try me sometimes, the constant attacks have ceased. I am led of the Holy Spirit in translation, intercession, spiritual warfare, and discernment, to be proactive in exposing, contending against, dismantling, and overthrowing these demonic forces, systems, and workers. I have also been given favor with God to translate at will to engage in kingdom purposes in spiritual realms. I am also able to bring deliverance to people who illegally astral project, teach people how to deal with spiritual warfare in these realms, and equip those who are called to translate and operate in spiritual realms.

Please note that people will come to ministry events, counseling, deliverance and need help closing doors to astral projection that they are experiencing because of generational bondage or because of their own sins and open doors. It will be important to decipher if they have a gift and calling to these realms and regions, so that you close illegal doors, but preserve and identify their gifts and callings while also training them how to be guided by the Holy Spirit when God requires them to translate in the spirit. When I was seeking deliverance, believers contended I had personal sin issues. Their condemnations and prayers caused me increased warfare as the things they prayed and spoke over me were used as open doors for demons and wicked astral projectors to further attack me. They did not discern that I had a gift and calling in this area nor that I had generational doors that were taking advantage of my gift and calling. It would be years later after experiencing constant attacks, that I stumbled across a seasoned balanced deliverance minister, who identified my gift and gave me revelation to seek God for how to walk in it. As I used this revelation, I was able to hear God about my generational line, close illegal doors where necessary, and seek Holy Spirit about how to walk in my gifts and calling in this area. Be open to moving beyond your religious familiarity by focusing on the correct specific key to break a person free. People do not need what you can handle about them or what religion has taught you. They need what God is speaking to you about them and what Holy Spirit is teaching you about them.

Closing doors to incubus, succubus, astral projectors:

- Repent for personal sins and transgressions.
- Repent for generational sins, curses, dedications, and covenants made to devils; repent for generational witchcraft practices, activities upon trading floors in demonic kingdoms, exchanges done on altars and demonic trading floors. Also break soul ties between the person and these generational trading floors, altars, and the demonic kingdoms tied to them. Search out insight in this area from Holy Spirit and close doors that he reveals.
- Break curses, covenants, and dedications made to demons and idols.
- Break soul ties and claims demons and their demonic kingdom have laid hold on the person.
- Spend time healing their soul, heart, mind, and body of any trauma areas.
- Shut demonic portals in the specific areas that apply to the person.
- Have the person confess Jesus Christ is Lord of their lives and lead them in a confession to rededicate their life and soul back in right standing and covenant with God.
- Declare that Jesus Christ is Lord over their soul and command their soul come from under any other ungodly or worldly governance and jurisdiction and to be reunited with him.
- Command their soul to eternally remain in their body and to be subject to the Holy Spirit on the inside of them.
- Spend time praising and worshipping and soaking the soul in the excellency of God.

- If you have a calling as an apostle, prophet, seer, or intercessor who scouts out spiritual realms for intel, purchase my book, *"Ascending Into Heavenly Realms,"* to learn how to legally translate and operate in this area of your calling.

Depending on the reason for astral projection, some people may require a season of contending for breakthrough for their souls. As they or someone in their generational line could have been chosen to serve or reign in the demonic realm, could have engaged in high level witchcraft practices, or could have pledged to high level rankings in Greek organizations. These idol gods will sometimes war for their souls, so, a season of spiritual warfare may be needed before deliverance and healing can fully manifest. I had to war for over a year for my breakthrough. Demons were enraged that I wanted to serve God and not the devil.

RESOURCES

Books by Dr. Taquetta Baker

Ascending Into Heavenly Realms By Taquetta Baker
Let Their Be Sight By Taquetta Baker
Unmasking The Power Of The Scouts Volume II: Soul Stealers

Websites

Blueletterbible.com
Biblestudytools.com
Dictionary.com
Olivetree.com
Strong's Exhaustive Bible Concordance Online Bible Study Tools

Shift right now!

Be a Healthy You!

Kingdom Shifters Product Line

Products available at kingdomshiftingbooks.com and amazon.com	
Books (Paperback, Kindle, and e-books available)	
Healing the Wounded Leader	There is an App for That
Apostolic Governing	Sustaining The Vision Workbook
Apostolic Mantle	Annihilating Church Hurt
Healing the Wounded Leader	Discerning the Voice of God
Release the Vision	Feasting in His Presence
Birthing Books That Shift Generations	Prayers that Shift Atmospheres
Atmosphere Changes (Weaponry)	Dismantling Homosexuality
Strategies for Eradicating Racism	Let There Be Sight
Kingdom Shifters Decree That Thang	Kingdom Watchman Builder on the Wall
Kingdom Heirs Decree That Thang	Kingdom Keys to Governing Relationships
Fivefold Operations – Manuals I, II, and III	Unmasking the Power of the Scouts – Volumes I and II
Processing Grief & Loss	Cultivating Destiny From The Womb
Kingdom Wellness Counseling & Mentoring Manual I	Deliverance From The Stronghold of Suicide
Truth About Willful Sin	Ascending Into Heavenly Realms
Gatekeeping Regions For God's Glory	KW Life Coaching Manual
Books for Liturgical / Interpretive Dance Ministries	
Dance & Fivefold Ministry	Dance from Heaven to Earth
Spirits that Attack Dance Ministers	Dancers! Dancers! Dancers! Decree That Thang
CD's	
Decree That Thang	Kingdom Heirs Decree That Thang
Teaching and Worship	

www.ingramcontent.com/pod-product-compliance
Lightning Source LLC
Chambersburg PA
CBHW081232020426

42331CB00012B/3135